Leckie
the education publisher
for Scotland

National 5 Physics Lab Skills

for SQA assessment

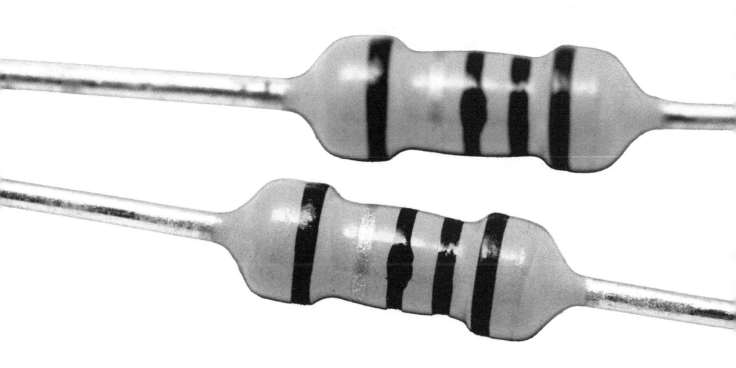

Michael Murray

© 2019 Leckie

001/24012019

10 9 8 7 6 5 4 3

ISBN 9780008329655

Published by
Leckie
An imprint of HarperCollins Publishers
Westerhill Road, Bishopbriggs, Glasgow, G64 2QT
T: 0844 576 8126 F: 0844 576 8131

leckiescotland@harpercollins.co.uk   www.leckiescotland.co.uk

HarperCollins Publishers
Macken House, 39/40 Mayor Street Upper, Dublin 1 D01 C9W8 Ireland

A CIP Catalogue record for this book is available from the British Library.

Publisher: Sarah Mitchel
Commissioning editor: Gillian Bowman
Managing editor: Craig Balfour

Special thanks to
Jouve India (layout, illustration and project management)
Dr Jan Schubert (proofreading)
Jess White (copyediting)

Contents

Underlying physics

These introductory statements cover the underlying physics which is needed to understand the context of the experiment and could help with any Assignment based on it.

Learning outcomes

This is a summary of the skills you will develop during each experiment.

Aim

A clear statement of the aim of the experiment is given. Any conclusion must be based on this aim

Apparatus list

Your teacher will ensure that all the apparatus you need for the experiment can be found in the laboratory. You can use this list to check that you have everything you need to start your work.

Safety notes

You should be aware of safety when doing an experiment. These notes will help you be aware of any safety issues! Your teacher will advise on safety information for each experiment, so pay attention.

Precautions

We've included some precautions which will help to make your experiment more accurate, valid or reliable.

Method

Always make sure you read every step of the method before you begin work. This will help you avoid mistakes and will give you an idea of what outcomes to look for as you complete each step.

Record your results

For each experiment there is a place to record the results of your work. Make sure you keep your data tables, graphs, answers and calculations clear and neat.

Check your understanding and Exam-style questions

For each experiment, there are questions designed to check your understanding of the work you've just completed. There is also an exam-style questions to help you prepare you for exam questions based on experimental and scientific inquiry skills.

Assignment support

For your assignment, you must carry out an experiment and collect data for use in your report. This section gives you some ideas of experiments you could carry out for your assignment.

1 Verifying Ohm's law

Underlying physics

Some electrical components have a constant resistance. If we double the potential difference, the current doubles. We call these components **ohmic conductors**. Other components do not have a constant resistance – increasing the potential difference might alter the current but it does not change proportionately. These are called **non-ohmic conductors**. We can tell which are which by testing them, plotting graphs of the data and analysing the shape of the graph.

Learning outcomes

- Take measurements using ammeters and voltmeters.
- Process information by calculating averages.
- Present information in a scatter graph.
- Interpret graphs that represent direct proportion.
- Draw valid conclusions from graphs.
- Calculate the gradient of the line of best fit.

Aim

To verify the relationship between potential difference and current for a fixed value of resistance.

Apparatus list

- 0–12 V variable power supply
- voltmeter or multimeter to measure voltage
- ammeter or multimeter to measure current (you may need a meter capable of measuring in the milliamp range)
- resistor (for example 5 Ω, 1 W)

Safety notes

- Keep electrical equipment away from water.
- Some components may get hot so take care when handling circuitry.
- Check the equipment before use. If it appears damaged, do not use it.

Experimental precautions

- Use a direct current source, not an alternating current source.
- Use low values of potential difference.
- Only switch on the power supply to take readings from the meters, then switch off.

Method

1. Construct the circuit as shown in **Figure 1.1**.

2. By adjusting the variable power supply, alter the current in even steps (for example 0·20 A, 0·40 A, 0·60 A, etc.) and record the values of current and potential difference in **Table 1.1**.

3. Take five pairs of readings. Turn off the power supply between readings to ensure the temperature of the resistor is kept constant.

4. Repeat measurements and calculate an average potential difference for each current value.

5. Plot a scatter graph of current against average (potential difference) and draw the line of best fit.

Figure 1.1

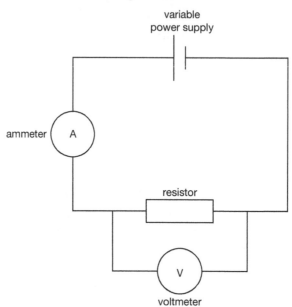

Record your results

Table 1.1

Current (A)	Potential difference (V)			
	1	**2**	**3**	**Average**
0·00	0·00	0·00	0·00	0·00

Graph 1.1

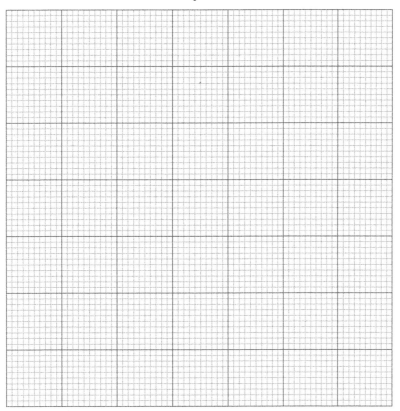

Check your understanding

1. Analyse your graph and describe the relationship between potential difference and current for a fixed resistor at a constant temperature. [2]

   ....

   ....

   ....

2. Explain why the power supply should be turned off between readings. [2]

   ....

   ....

   ....

3. Determine the resistance of your resistor by calculating the gradient of your line of best fit. Make sure you choose two sets of points that are on the line of best fit you have drawn. [2]

   ....

   ....

   ....

4. Describe how you can improve the accuracy of this calculated result. [1]

   ....

   ....

Exam-style question

A student investigates the resistance of a resistor using the circuit shown below.

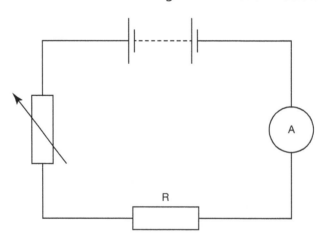

a. Complete the circuit diagram to show where a voltmeter should be connected to measure the potential difference across resistor R. [1]

b. Explain how the student can use the circuit to obtain a range of values for potential difference and current. [1]

....

....

c. The results of the student's experiment are shown below.

Current in resistor R (A)	Potential difference across resistor R (V)
0·20	1·00
0·44	2·20
0·87	4·35
1·24	6·20

Use all the data to determine the resistance of resistor R. [4]

....

....

....

....

....

Assignment support

You could use this technique to generate data for your assignment.

Alternatively, a filament bulb could be used in place of the fixed resistor. This would allow for the investigation of the relationship between current and potential difference for a non-ohmic conductor.

2 Verifying the pressure–volume law (Boyle's law)

Underlying physics

There are three laws that link the behaviour of gases. Each of these laws considers three properties of gases: pressure (p), temperature (T) and volume (V). The volume of a gas is the volume of the container that is holding the gas. The pressure of a gas is caused by particles colliding with the walls of the container. The pressure can be increased if the particles collide with the container walls more frequently, or if the particles collide with the container walls with greater force. The temperature of a gas depends on the kinetic energy of the gas particles. The pressure–volume law, or Boyle's law, considers the relationship between the pressure and volume of a fixed mass of gas at a constant temperature.

Learning outcomes

- Make accurate measurements of volume and pressure.
- Present information in a scatter graph.
- Interpret graphs that represent direct proportion.
- Draw valid conclusions from graphs.

Aim

To investigate how the pressure of a gas changes when the volume is changed.

Apparatus list

- trapped air column (oil column inside glass tube)
- volume scale
- pump
- pressure gauge
- tap between pump and pressure gauge

Safety notes

- Do not increase the pressure of the trapped air to unsafe levels.

Experimental precautions

- Check all seals before starting the experiment.

Method

1. Set up the apparatus as shown in **Figure 2.1**.
2. Connect the pump to the apparatus and use it to increase the pressure of the trapped air.
3. Seal the apparatus using the tap.
4. Record values for volume and pressure.
5. Open the tap to reduce the pressure by a small amount and then reseal.
6. Allow the oil to settle
7. Open the tap again, slightly, and adjust the oil level until it lies on an easily read value.
8. Record values for volume and pressure.
9. Repeat measurements and enter your values in **Table 2.1**.
10. Plot a graph of pressure against volume in **Graph 2.1**.
11. Plot a graph of pressure against 1/volume in **Graph 2.2**.
12. Use the graphs to determine the relationship between the pressure and volume of a gas.

Figure 2.1

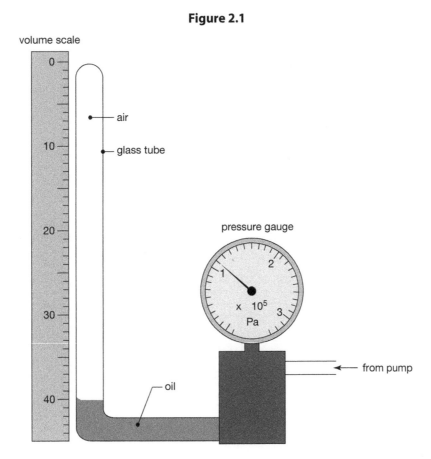

Record your results

Table 2.1

Pressure (Pa)	Volume (units)

Graph 2.1

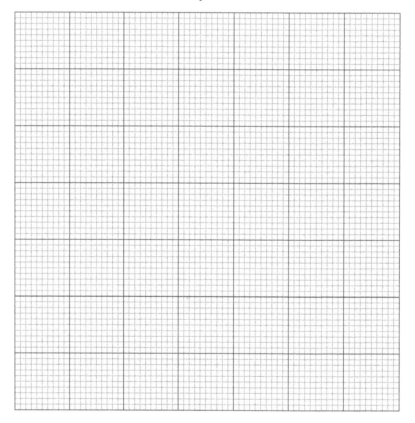

Graph 2.2

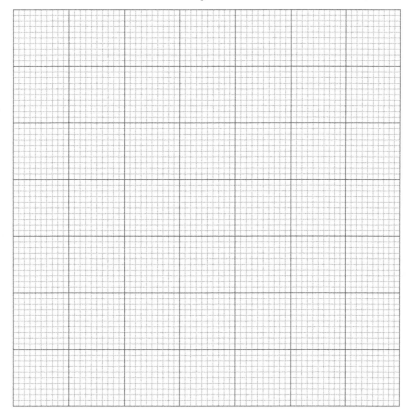

Check your understanding

1. Write a conclusion based on the aim of the experiment. [1]

....

....

2. Identify the relationship between the pressure and volume of a gas. [3]

....

....

....

....

3. Explain the effect of changing volume on pressure using the kinetic model. [2]

....

....

....

Exam-style question

A high-volume bicycle pump contains 20 cm³ of air at a pressure of 1.0×10^5 Pa.

The piston of the pump is pushed inwards until the volume of air is 8.4 cm³.

The temperature of the air inside the pump remains constant.

a. Calculate the final pressure of the air inside the pump. [3]

....

....

....

....

b. Using the kinetic model, explain what happens to the pressure of the air inside the pump as the volume decreases. [3]

....

....

....

....

Assignment support

You could use this technique to generate data for your assignment.

A 20 cm³ air syringe can be used in place of the apparatus used in this experiment. The data can be further extended through calculations showing that pV = constant.

3 Verifying the pressure–temperature law

Underlying physics

There are three laws that link the behaviour of gases. Each of these laws considers three properties of gases: pressure (p), temperature (T) and volume (V). The volume of a gas is the volume of the container that is holding the gas. The pressure of a gas is caused by particles colliding with the walls of the container. The pressure can be increased if the particles collide with the container walls more frequently, or if the particles collide with the container walls with greater force. The temperature of a gas depends on the average kinetic energy of the gas particles. The pressure–temperature law considers the relationship between the pressure and temperature of a fixed mass of gas at a constant volume.

Learning outcomes

- Make accurate measurements of pressure and temperature.
- Convert degrees Celsius to Kelvin.
- Present information in a scatter graph.
- Interpret graphs that represent direct proportion.
- Draw valid conclusions from graphs.
- Evaluate experimental procedures and suggest improvements.

Aim

To investigate how the pressure of a gas changes when the temperature is changed.

Apparatus list

- rigid flask
- water bath
- thermometer
- pressure gauge
- short length of rubber tubing

Safety notes

- The water in the water bath can be extremely hot so handle with care.
- Keep water bath away from mains electrical outlets.

Experimental precautions

- Check all seals before starting the experiment.

Method

1. Set up the apparatus as shown in **Figure 3.1**.

2. Heat the water in the water bath to the starting temperature, for example 90°C.

3. Record values of temperature and pressure in **Table 3.1**.

4. Record the pressure at different temperatures as the water cools.

5. Plot a graph of pressure against temperature (°C) in **Graph 3.1** (like the one shown here). Include negative values on the *x*-axis to −300°C. Then extend the line of best fit to find the temperature at which the pressure is 0 Pa.

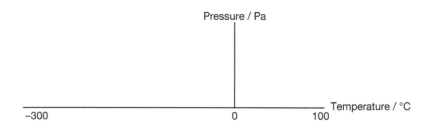

6. Convert the temperature to kelvin by adding 273 to the degrees Celsius reading and record in **Table 3.2**.

7. Plot a graph of pressure against temperature in kelvin in **Graph 3.2**.

8. Use the graph to determine the relationship between the pressure and temperature of a gas.

Figure 3.1

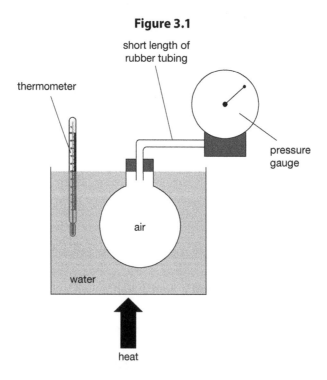

Record your results

Table 3.1

Temperature (°C)	Pressure (Pa)

Graph 3.1

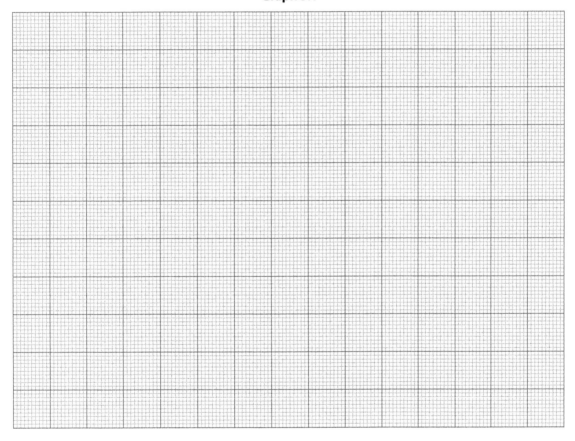

Table 3.2

Temperature (K)

Graph 3.2

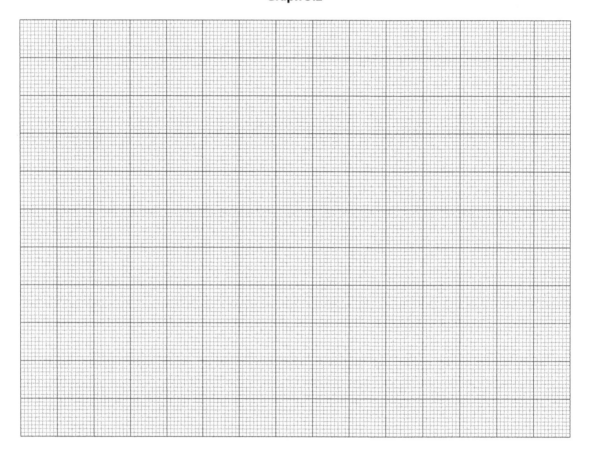

Check your understanding

1. Write a conclusion based on the aim of the experiment. [1]

   ....

   ....

2. Identify the relationship between the pressure and temperature of a gas. [3]

   ....

   ....

   ....

   ....

3. Explain how the apparatus could be improved to make the experiment more accurate. [2]

   ....

   ....

   ....

Exam-style question

An experiment is set up to investigate the relationship between the pressure and temperature of a fixed mass of gas as shown.

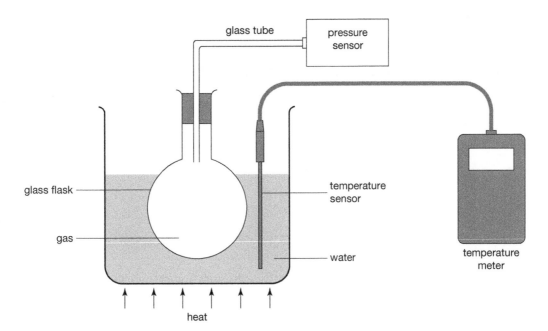

The following readings are recorded.

Pressure (kPa)	101	108	115	120
Temperature (K)	295	315	335	350

    a. Using **all** the data, determine the relationship between the pressure and temperature of the gas. [3]

....

....

....

....

    b. Using the kinetic model, explain why the pressure of a gas increases as its temperature increases. [3]

....

....

....

....

Assignment support

You could use this technique to generate data for your assignment.

The data can be further extended through calculations showing that $\frac{p}{T}$ = constant.

4 Verifying the volume–temperature law (Charles' law)

Underlying physics

There are three laws that link the behaviour of gases. Each of these laws considers three properties of gases: pressure (p), temperature (T) and volume (V). The volume of a gas is the volume of the container that is holding the gas. The pressure of a gas is caused by particles colliding with the walls of the container. The pressure can be increased if the particles collide with the container walls more frequently, or if the particles collide with the container walls with greater force. The temperature of a gas depends on the average kinetic energy of the gas particles. The volume–temperature law considers the relationship between the volume and temperature of a fixed mass of gas at a constant pressure.

Learning outcomes

- Make accurate measurements of volume and temperature.
- Convert degrees Celsius to Kelvin.
- Present information in a scatter graph.
- Interpret graphs that represent direct proportion.
- Draw valid conclusions from graphs.

Aim

To investigate how the volume of a gas changes when the temperature is changed.

Apparatus list

- water bath
- thermometer
- capillary tube with attached scale to measure volume
- bead of sulfuric acid, mercury or similar

Safety notes

- Care must be taken when handling even small amounts of mercury. Any breakages should be reported immediately.
- The water in the water bath can be extremely hot so handle with care.
- Keep water bath away from mains electrical outlets.
- Be extra careful with the capillary tube if it contains concentrated sulfuric acid.

Experimental precautions

- After continued use, the beads can sometimes separate. This can result in incorrect volume readings. A thin piece of wire can be inserted into the capillary tube to push the bead back together.

Method

1. Set up the apparatus as shown in **Figure 4.1**.
2. Heat the water to the starting temperature, for example 90°C.
3. Record values of volume and temperature in **Table 4.1**.
4. Convert the temperature to kelvin by adding 273 to the degrees Celsius reading.
5. Record the volume at different temperatures as the water cools.
6. Plot a graph of pressure against temperature in kelvin in **Graph 4.1**.
7. Use the graph to determine the relationship between the volume and temperature of a gas.

Figure 4.1

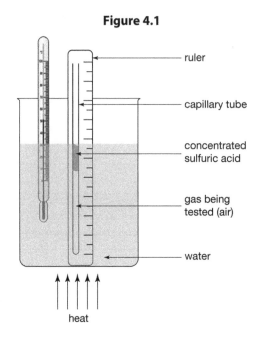

ruler

capillary tube

concentrated sulfuric acid

gas being tested (air)

water

heat

Record your results

Table 4.1

Temperature (°C)	Temperature (K)	Volume ( )

Graph 4.1

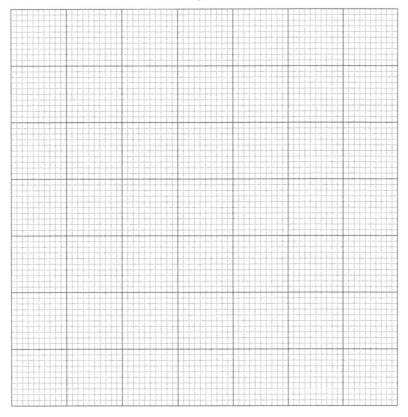

Check your understanding

1. Write a conclusion based on the aim of the experiment. [1]

....

....

2. Identify the relationship between the volume and temperature of a gas. [3]

....

....

....

....

3. Explain, using the kinetic model, the effect of changing temperature on the volume of a gas. [3]

....

....

....

....

Exam-style question

A volume of a fixed mass of gas is measured at five different temperatures. The pressure of the gas is kept constant throughout. The results are shown in the table below.

Volume (cm³)	0·200	0·207	0·214	0·221	0·228
Temperature (K)	280	290	300	310	320

  a. Using **all** the data, establish the relationship between the volume and the Kelvin temperature of the gas. [3]

....

....

....

....

  b. Calculate the volume of the gas when the temperature is 62°C. [3]

....

....

....

....

Assignment support

You could use this technique to generate data for your assignment.

The data can be further extended through calculations showing that $\frac{V}{T}$ = constant.

5 Measuring the half-life of a radioactive source

Underlying physics

Radioactive sources contain unstable isotopes that decay over time until they become stable. Therefore, the activity of a radioactive source decreases with time. The emission of a radioactive particle is a random process, in that we cannot predict precisely when a nucleus will decay. We can, however, make predictions as to how many nuclei will decay in a certain period of time. The half-life is the time taken for the activity to fall to half its initial value. Different radioactive sources have different half-lives. For example, uranium-235 has a half-life of 703·8 million years whereas protactinium-234 has a half-life of around 70 seconds.

Learning outcomes

- Make accurate measurements of activity and time.
- Calculate corrected count rate.
- Present information in a scatter graph.
- Draw valid conclusions from graphs.
- Determine the half-life of a radioactive source from an activity–time graph.

Aim

To determine the half-life of a radioactive source.

Apparatus list

- Geiger–Müller tube
- counter/ratemeter
- timer
- radioactive source, for example protactinium-234 generator

Safety notes

- Handle the radioactive source with tongs.
- Direct the source away from the body.
- Use a perspex safety screen.

Experimental precautions

- This experiment should only be carried out under the careful supervision of a teacher. Students under the age of 16 will not participate directly in the experiment.

Method

1. Ensure the radioactive source is several metres away from the Geiger-Muller tube. Start the counter and timer at the same time.
2. Record the background count rate for 1 minute.
3. Determine the background count rate in counts per second.
4. Position the Geiger–Müller tube so that it is facing towards the radioactive source.
5. Start the counter and timer at the same time.
6. Record the count rate every 10 seconds for around 5 minutes in **Table 5.1**.
7. Subtract the background count rate from all values to determine the corrected count rate.
8. Plot a graph of corrected count rate against time in **Graph 5.1** and draw a best fit curve.

Record your results

Background count rate for 1 minute = ... counts per minute

Background count rate = ... counts per second

Table 5.1

Time (s)	Count rate (counts per second)	Corrected count rate (counts/ second)
0		
10		
20		
30		

Graph 5.1

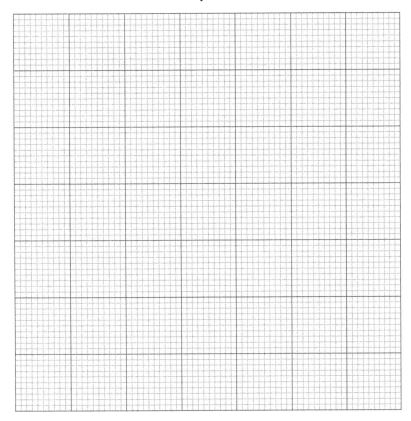

Check your understanding

1. From the graph of count rate against time, what happens to the activity with time? [1]

....

....

2. Use the graph to determine the half-life of the radioactive source. [1]

....

....

3. What conclusion can be made from the shape of the graph? [1]

....

....

Exam-style question

An experiment is carried out to determine the half-life of a radioactive source using the apparatus shown.

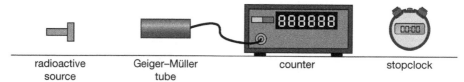

radioactive   Geiger–Müller   counter   stopclock
source   tube

a. State what is meant by the term *half-life*. [1]

....

....

b. The data from the experiment is shown in the graph.

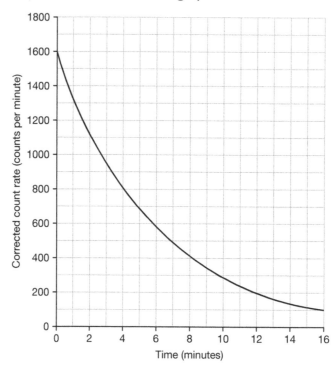

i. Describe how the apparatus could be used to obtain the data shown in the graph. [3]

....

....

....

....

ii. Use information from the graph to determine the half-life of the radioactive source. [1]

....

....

Assignment support

You could use this technique to generate data for your assignment.

The experiment could be repeated to obtain an average half-life.

6 Measuring the average speed of an object

Underlying physics

Speed is defined as the distance travelled in a unit of time and is measured in metres per second ($m\ s^{-1}$). The average speed of an object is defined as the distance for the whole journey divided by the total time taken. Average speed can be calculated using the relationship $\bar{v} = \dfrac{d}{t}$, where $\bar{v}$ is the average speed measured in metres per second, d is the distance for the whole journey measured in metres and t is the total time taken measured in seconds.

Learning outcomes

- Make accurate measurements of distance.
- Carry out calculations involving average speed, distance and time.
- Present information in a scatter graph.
- Draw valid conclusions from graphs.

Aim

To determine the average speed of a trolley.

Apparatus list

- mask
- trolley
- runway
- two light gates
- electronic timer
- measuring tape

Safety notes

- Prevent the trolley from falling off the bench/table.

Experimental precautions

- Adjust the light gates to the correct height to ensure they do not collide with the trolley.
- Reset the timer between measurements.

Method

1. Set up the apparatus as shown in **Figure 6.1**.
2. Place the runway on a level surface and raise one end by several centimetres.
3. Mark a midpoint on the centre of the track.
4. Mark a starting point for the trolley.
5. Place both light gates at a set distance from the midpoint, for example 30 cm.
6. Release the trolley from the top of the slope.
7. Record values for time and distance (*d*) in **Table 6.1**.
8. Calculate the average speed using the relationship:

   average speed = distance between the light gates (*d*) ÷ time taken
9. Repeat for new values of distance by decreasing the distance (*d*) between the light gates. Make sure the light gates are equidistant from the midpoint of the slope.
10. Plot a graph of average speed against distance in **Graph 6.1**.

Figure 6.1

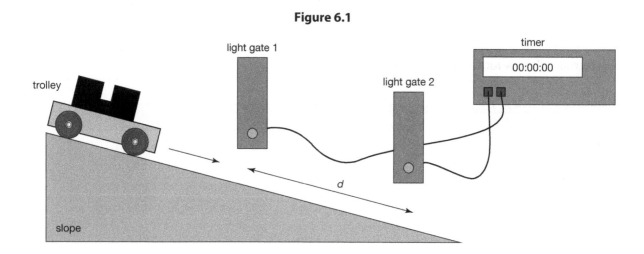

Record your results

Table 6.1

Distance (m)	Time (s)	Average speed (m s^{-1})

Graph 6.1

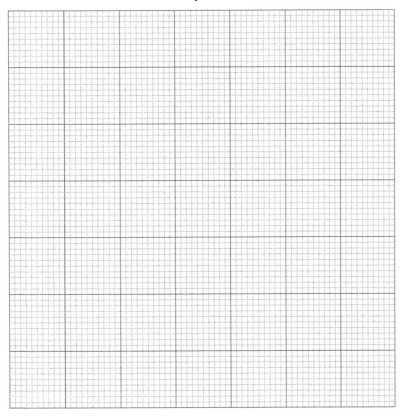

Check your understanding

1. State the independent and dependent variables in this experiment. [2]

   Independent variable:

   ....

   Dependent variable:

   ....

2. List two variables that were controlled in the experiment. [2]

   ....

   ....

   ....

Exam-style question

A bungee jumper wants to calculate their average speed during a jump.

On the first descent they fall 58 metres in 7·25 seconds.

   a. What device can be used to measure the time of the descent? [1]

   ....

   ....

   b. Calculate the average speed during the descent. [3]

   ....

   ....

   ....

   ....

Assignment opportunity

You could use this technique to generate data for your assignment.

The average speed of an object can also be determined by using a stopwatch to determine the time taken for the trolley to cover a measured distance. However, this method is less reliable than using electronic timing methods.

7 Measuring the instantaneous speed of an object

Underlying physics

Speed is defined as the distance travelled in a unit of time and is measured in metres per second (m s^{-1}). The instantaneous speed of an object is its speed at any given instant in time. Instantaneous speed can be calculated using the relationship $v = \dfrac{d}{t}$, where v is the instantaneous speed measured in metres per second, d is the distance travelled in metres and t is the time taken in seconds. In order to measure instantaneous speed, we need to measure very short distances covered in very short periods of time.

Learning outcomes

- Make accurate measurements of length and time.
- Carry out calculations involving average speed, distance and time.
- Process information by calculating averages.

Aim

To determine the instantaneous speed of a trolley.

Apparatus list

- trolley
- runway
- a mask of a measured length
- light gate
- electronic timer

Safety notes

- Prevent the trolley from falling off the bench/table.

Experimental precautions

- Adjust the light gates to the correct height to ensure they do not collide with the trolley.
- Reset the timer between measurements.

Method

1. Set up the apparatus as shown in **Figure 7.1**.
2. Place the runway on a level surface and raise one end by several centimetres.
3. Attach a mask to the trolley that is of a measured length, for example 5 cm.
4. Record the length of the mask.
5. Mark a starting point for the trolley.
6. Place the light gate at a position towards the bottom of the slope.
7. Release the trolley from the top of the slope.
8. Record a value for the time taken for the mask to pass through the light gate beam in **Table 7.1**.
9. Repeat measurements twice more. Calculate the average time taken.
10. Calculate instantaneous speed using the relationship:

$$\text{instantaneous speed} = \frac{\text{length of card}}{\text{average time}}$$

Figure 7.1

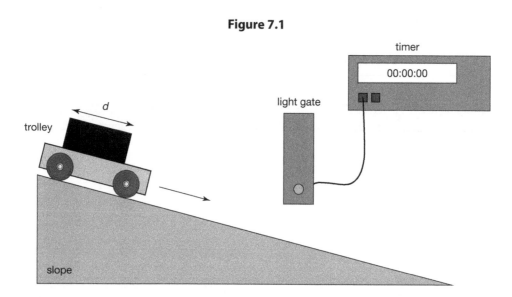

Record your results

Length of mask = ... m

Table 7.1

Attempt	Time (s)
1	
2	
3	

Average time = ... s

Calculation:

$$v = \frac{d}{t} = ...$$

Check your understanding

1. Describe the difference between average and instantaneous speed.   [2]

   ....

   ....

   ....

2. Suggest **two** ways in which the experiment is carried out that could lead to inaccuracies in the final calculated value of instantaneous speed.   [2]

   ....

   ....

   ....

3. Explain why it is not possible to measure the time interval in this experiment accurately by using a manual stopclock.   [2]

   ....

   ....

   ....

Exam-style question

The apparatus shown below is used to determine the acceleration of a trolley.

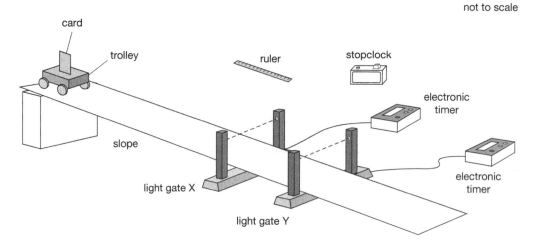

not to scale

card
trolley
ruler
stopclock
electronic timer
electronic timer
slope
light gate X
light gate Y

Some of the measurements taken are shown.

Time for the card to pass through light gate Y	0·042 s
Distance between light gate X and light gate Y	0·36 m
Length of the card	0·050 m
Time for the trolley to pass between light gate X and light gate Y	0·38 s

a. State the additional measurement required to determine the instantaneous speed of the trolley at light gate X. [1]

....

....

b. Calculate the instantaneous speed of the trolley at light gate Y. [3]

....

....

....

....

Assignment support

You could use this technique to generate data for your assignment.

The instantaneous speed of an object could be further investigated by varying the angle of the slope and determining the relationship between the angle of the slope and the instantaneous speed of the trolley.

Underlying physics

When an object changes speed (or velocity), it is said to be accelerating. Acceleration is a vector quantity and therefore has both size and direction. The direction of the acceleration is indicated by a positive or negative sign in the final value; if the acceleration is positive then the object is increasing speed, if the acceleration is negative then the object is decreasing speed. The definition of acceleration is the change in speed (or velocity) per second and is measured in metres per second squared (m s^{-2}).

The acceleration of an object can be calculated using the formula $a = \dfrac{v - u}{t}$, where a is the acceleration in m s^{-2}, v is the final speed in m s^{-1}, u is the initial speed in m s^{-1} and t is the time taken in seconds.

Learning outcomes

- Make accurate measurements of length and time.
- Carry out calculations involving acceleration, final speed, initial speed and time.

Aim

To determine the acceleration of a trolley.

Apparatus list

- trolley
- runaway
- double mask of a measured length (l)
- light gate
- electronic timer

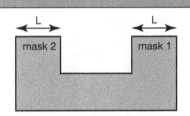

Safety notes

- Prevent the trolley from falling off the bench/table.

Experimental precautions

- Adjust the light gates to the correct height to ensure they do not collide with the trolley.
- Reset the timer between measurements.

Method

1. Set up the apparatus as shown in **Figure 8.1**.

2. Place the slope on a level surface and raise one end by several centimetres.

3. Attach a double mask that is of a measured length, for example 5 cm, to the trolley. Both parts of the mask must be the same length.

4. Record the length of the mask.

5. Mark a starting point for the trolley.

6. Place the light gate at a position towards the bottom of the slope.

7. Release the trolley from the top of the slope.

8. Use the electronic timer to record the following times:

   - time for mask 1 to pass through the light gate – this can be used to calculate the initial speed, u
   - time for mask 2 to pass through the light gate – this can be used to calculate the final speed, v
   - time for both parts of the double mask to pass through the light gate – this is time, t.

9. Calculate the acceleration using the relationship:

$$a = \frac{v - u}{t}$$

Figure 8.1

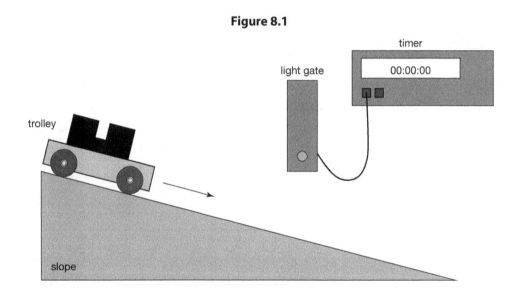

Record your results

Length of mask = ... m

Time for mask 1 to pass through light gate = ... s

Time for mask 2 to pass through light gate = ... s

Time for double mask to pass through light gate = ... s

Calculation of initial speed, u:

$$u = \frac{\text{length of mask}}{\text{time for mask 1 to pass through light gate}} = \text{................................} \ \text{m s}^{-1}$$

Calculation of final speed, v:

$$v = \frac{\text{length of mask}}{\text{time for mask 2 to pass through light gate}} = \text{................................} \ \text{m s}^{-1}$$

Calculation of acceleration, a:

$$a = \frac{v - u}{t} = \text{................................} \ \text{m s}^{-2}$$

Check your understanding

1. Explain why a double mask is used to determine acceleration in this experiment [2]

    ....

    ....

    ....

2. List **two** variables that are controlled in this experiment. [2]

    ....

    ....

    ....

Exam-style question

A cyclist is found to have an acceleration of 1.5 m s^{-2}.

a. State what is meant by an acceleration of 1.5 m s^{-2}. [1]

....

....

b. The cyclist approaches a set of traffic lights at 16 m s^{-1}. The lights turn red and the cyclist applies the brakes, coming to rest after 4.8 s.

Calculate the acceleration of the cyclist while braking. [3]

....

....

....

....

Assignment support

You could use this technique to generate data for your assignment.

Acceleration data can also be generated using two light gates and a single mask. This could be further extended to examine the acceleration at different points on the slope.

Answers

1 Verifying Ohm's law

Check your understanding

1. The current in a resistor is directly proportional to the voltage across it. As current increases, resistance stays the same since the gradient of the line is the same at all points. [2]
2. To keep the temperature of the resistor constant [1]
   Resistance increases as temperature increases [1]
3. Correct substitution in gradient formula [1]
   Final answer, including units [1]
4. Use lower current values to reduce the amount of energy lost as heat energy.
   OR
   Use a large wattage resistor (at least 2 W) to reduce heating effects caused by the current. [1]

Exam-style question

a. Voltmeter across resistor R [1]
b. Change the resistance of the variable resistor [1]
c. Ohm's law stated [1]
   All correct substitutions shown [2]
   $5\,\Omega$ [1]

2 Verifying the pressure–volume law (Boyle's law)

Check your understanding

1. As the volume (of air) increases, pressure decreases
   As volume decreases, pressure increases [1]
2. $p \propto \dfrac{1}{V}$ [1]
   $pV = $ constant [1]
   $p_1 V_1 = p_2 V_2$ [1]
3. Decreasing volume causes particles to collide with container walls more frequently [1]
   More frequent collisions increase the pressure of the gas [1]

Exam-style question

a. $p_1 V_1 = p_2 V_2$ [1]
   $1 \times 10^5 \times 20 = p_2 \times 8{\cdot}4$ [1]
   $2{\cdot}4 \times 10^5$ Pa [1]
b. Particles collide with container walls more frequently [1]
   Force is greater [1]
   Pressure increases [1]

3 Verifying the pressure–temperature law

Check your understanding

1. As the temperature decreases, the pressure decreases [1]
2. $p \propto T(k)$ [1]
   $\dfrac{p}{T} = $ constant [1]
   $\dfrac{p_1}{T_1} = \dfrac{p_2}{T_2}$ [1]

3. Place thermometer inside the flask [1]
   This will measure the temperature of the gas, not the water surrounding the flask [1]

Exam-style question

a. All four substitutions [1]
   All values correct (342, 343, 343, 343) [1]
   $\dfrac{p}{T(k)} = $ constant [1]
b. Increase in temperature increases average kinetic energy of particles [1]
   Particles hit the container walls more frequently [1]
   Particles hit the container walls with greater force causing the pressure to increase [1]

4 Verifying the volume–temperature law (Charles' law)

Check your understanding

1. As temperature increases, volume increases [1]
2. $V \propto T(k)$ [1]
   $\dfrac{V}{T} = $ constant [1]
   $\dfrac{V_1}{T_1} = \dfrac{V_2}{T_2}$ [1]

3. As the temperature decreases the average kinetic energy of the particles decreases. [1]
   Particles hit the container walls less frequently [1]
   Volume decreases to keep pressure constant [1]

Exam-style question

a. All five substitutions [1]
   $\dfrac{V}{T} = 7{\cdot}1 \times 10^{-4}$ [1]
   $\dfrac{V}{T(k)} = $ constant [1]
b. $\dfrac{V_1}{T_1} = \dfrac{V_2}{T_2}$ [1]
   $\dfrac{0{\cdot}2}{280} = \dfrac{V_2}{335}$ [1]
   $V_2 = 0{\cdot}240$ cm³ [1]

5 Measuring the half-life of a radioactive source

Check your understanding

1. Activity decreases [1]
2. Dependent on sources available. If a protactinium generator is used, then the half-life is approximately 70 s. [1]
3. Count rate always takes the same amount of time to half [1]

Exam-style question

a. Time taken for the activity/corrected count rate to half [1]

b. i. Measure the count in a set time interval [1]

Repeat at regular intervals [1]

Measure background count rate and subtract
to find corrected count rate [1]

ii. 4 minutes [1]

6 Measuring the average speed of an object

Check your understanding

1. Independent variable: distance travelled [1]

Dependent variable: average speed [1]

2. The starting point for the trolley [1]

The angle of the slope [1]

Exam-style question

a. Stopwatch [1]

b. Equation stated [1]

All correct substitutions shown [1]

8 m s^{-1} [1]

7 Measuring the instantaneous speed of an object

Check your understanding

1. Average speed is the average of all speeds across
an entire journey [1]

Instantaneous speed is the speed at a particular
moment in time [1]

2. The trolley is pushed rather than allowed to roll freely [1]

The length of the mask is measured incorrectly [1]

3. The time interval is very short [1]

Human reaction times are not sufficient to time it
accurately [1]

Exam-style question

a. Time for the card to pass through light gate X [1]

b. Equation stated [1]

All correct substitutions shown [1]

$1 \cdot 2 \text{ m s}^{-1}$ [1]

8 Measuring the acceleration of an object

Check your understanding

1. Two values of instantaneous speed are required to
calculate acceleration [1]

The instantaneous speed of mask 1 (initial speed)
and the instantaneous speed of mask 2 (final speed) [1]

2. Starting point of the trolley [1]

Method of release [1]

Exam-style question

a. The speed (or velocity) changes by $1 \cdot 5 \text{ m s}^{-1}$ every
second [1]

b. Equation stated [1]

All correct substitutions shown [1]

$-3 \cdot 3 \text{ m s}^{-2}$ [1]

42

Notes

Notes